Solar system
(The Sun and her Daughters)

Written by:
Emma Mahmoud Aljawarneh
Final Revision:
Dr. Bashar Beano

© 2023 by Emma Mahmoud Aljawarneh

All rights reserved. No part of this publication may be reproduced, distributed, or transmitted in any form or by any means without the prior written permission of the publisher.

ISBN: 979-8-336-10963-4

First Edition

Cover design by Ala'a Al Rashdan

Edited/Revised by Dr. Bashar Beano

The author has made every effort to ensure the accuracy of the information in this book. However, the information provided is for educational purposes only and does not constitute legal, financial, or professional advice.

Published by Dar AL-Ketab AL-Thaqafi
Irbid, Jordan

Printed in Jordan

Introduction

Astronomy is an ancient on the cutting edge, great discoveries were made centuries ago, also great discoveries are being made today. And great leaps forward in astronomical knowledge have often followed leaps forward in technology, the invention of telescope, the invention of the computer and so on.

When we think about our solar system, we usually assume that it has always been much as it is now, and always will be, but what we know of the solar system is only a mere snapshot in comparison to its 4.6 billion - year ago. However, even the earliest astronomers wanted to do more than predict the planets motions. They want to know what was really going on. When Copernicus, Galilio, Tycho Brahe, and Kepler finally succeeded in doing this quite well in the 16th and 17th centuries, it was a momentous time for astronomy and human understanding.

Understanding how the planets move is important, of course, but our understanding of the solar system hardly ends with that. In the last few decades of the 20th century and now into the 21th, astronomers have learned more about the solar system than in all the 400 years since planetary motions were pretty well nailed down. As this effort will show, the solar system family (The Sun and Her Daughters), which I hope to make an addition and enjoyment for the reader.

Solar system

The solar system that we live in has 8 planets including: Jupiter, Saturn, Uranus, Neptune, Earth, Venus, Mars and Mercury, but the ninth planet sometimes is Pluto. The solar system has 2 belts called Asteroid belt and Kuiper belt.

The first and the most important question, how does the solar system form?
The answer is in the Nebular theory, which tells us that our solar system was made up of a cloud of gas called nebula, located in the depths of the spiral arm of the Milky Way Galaxy that is formed by Big Bang. This huge cloud consists of two light elements (hydrogen, helium) and a little (oxygen, silicon, and iron).
The nebula rotates slowly around its center, which consists of a mass of complex vortices formed by what is known as attraction down. Under the influence of gravity, the nebula took the shape of the turntable with increasing heat and mass density at the center, which led to the sun formation.

The outer part of the largest nebula and the coolest, its materials such as water, ammonia and methane hardened and the union of materials of aluminum, iron, silicon and others with oxygen and crystallization at high temperatures formed dense rock material after a period became larger then become primitive planets.

When these planets move their orbits around the sun, it is able to grow by the attractive magnification that makes it sweep and collect many of the nearby materials from orbits to become major planets.

Have you've ever wondered what the worst way our solar system might end is?!

Our solar system might end like this:
1- A black hole entering our solar system and consumes everything.
2- The sun becoming a red giant and eat every single planet till Earth!!

Let's talk about some theories that have been asked a lot:

What will happen if a black hole enters our solar system?

Well that's a theory that has been going a lot, so let's answer it, well if a black hole enters our solar system with the same mass as the sun, we wouldn't be in trouble and let's say that it got replaced with our sun we would orbit the black hole how we used to orbit the sun, but if a black hole with huge mass like Sagittarius A* (one of the biggest black holes) enters our solar system, we would be doomed, why you're going to ask? Well our solar system would be swallowed from the Sun to the Kuiper belt.

What will happen to our solar system in the future?

First, in five billion years the sun is going to be red giant and eat every planet to the Earth, and in the next 100 years temperatures here on earth will get higher, the solar system might only survive for a few billion years before the sun stops burning.

Let's see how far every planet from the sun is: (AU means astronomical unit and it means the distance from the earth to the sun)

Mercury is 0.39 AU far from the sun which is pretty nice. However, Venus is 0.72 AU far from the sun pretty much close to the earth, the earth is the perfect number its 1 AU far from the sun, which is 150 million kilometers, Mars is good but not too far, its 1.52 AU far from the sun, kind of like earth. Now that we have finished the inner planets let's go to the outer planets. Jupiter is far and it's pretty far its 5.20 AU we jumped a lot from mars to Jupiter but Saturn is 9.54 AU we keep jumping a lot, but Uranus is 19.22 AU we jumped about ten AU, but last one on this list is Neptune its 30.06 AU far from the sun, that's pretty far.

Sphere	Mercury	Venus	Earth	Mars	Jupiter	Saturn	Uranus	Neptune
Mass	0.33	4.87	5.97	0.642	1898	568	86.8	102
Density	5429	5243	5514	3340	1326	687	1270	1638
Moons	0	0	1	2	95	146	27	14
Diameter	4,879	12,104	12,756	6,792	142,984	120,536	51,118	49,528

The Sun

The sun is the only star that we have in the solar system; in fact the sun has 99.8% of the solar system mass. The sun keeps every planet
in its orbit.

Sun facts:
The sun is an average size compered to others
The sun has the mass of 1.9891 × 1030 kilograms
The sun is the only and only star in our solar system
The sun produces 1038 neutrino every second

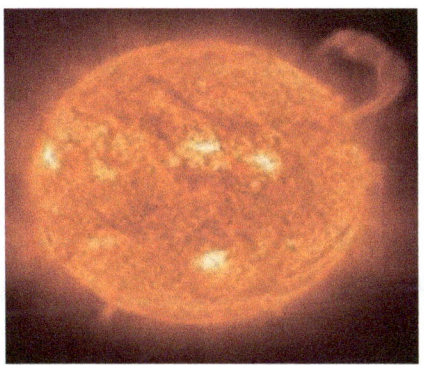

This picture of the sun was taken by a special filter that shows the complex magnetic field of our
beautiful sun.

How big is the sun?

The sun is literally humongous, in fact its 1,392,000 km or in a simplified way 109 earths across, we are just a grain of sand compared to it.

What are the layers of the sun?

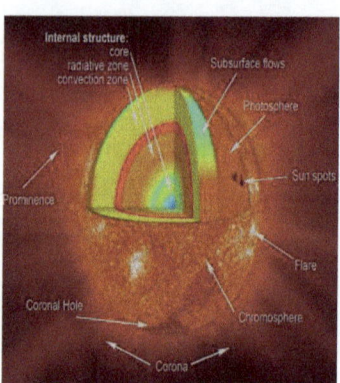

The inner layers are: Core, Radiative zone and convection zone, and the outer layers are:
Photosphere, the
Chromosphere, the
Transition Region and the Corona.

Do you know what sunspots are?

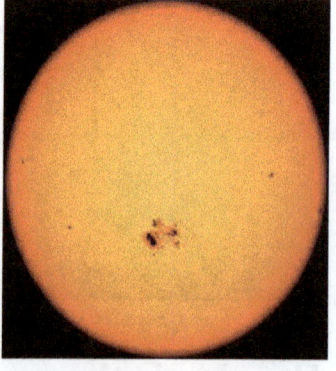

If you don't know sunspots, sunspots are areas of the sun that are slightly darker and cooler, **how do they appear?** They only appear at dark against the brightness of the rest of the surface of the sun. **But how do they occur?** They are caused by twisting and chaotic magnetic fields in the sun's convection zone.

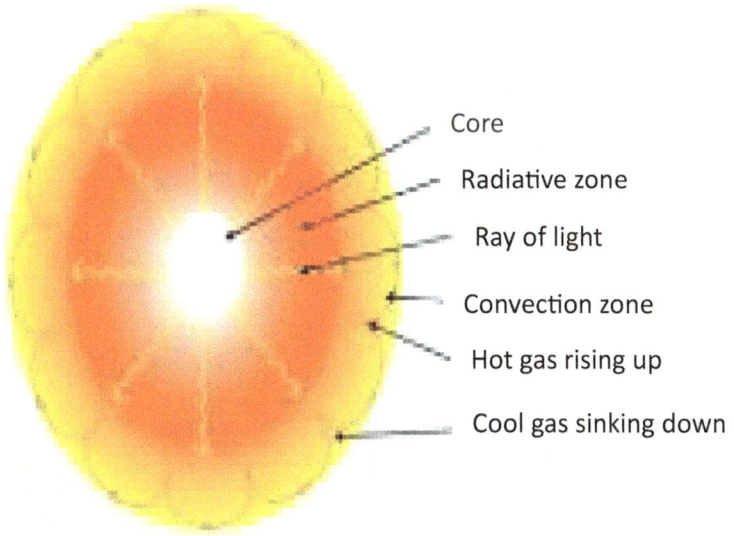

Core: the core is right in the middle of the sun, temperatures there are very high its 15 million Kelvin you might be sweating just reading this, this region of the sun is where nuclear reactions happen the most.

Radiative zone: well here the light, heat and X-rays that are produced fight their way so they can go to the surface. Actually, the ray of light over there will take 1 million years or more to get out. The region that saves energy for long time and it's the region that transfers the energy from core to the next layer by radiation. The temperature here is about 400,000 Kelvin.

Convection zone: the convection zone is an interesting zone, have you ever seen the air shimmer above the fire? You've probably been told by your parents or friends that it's because of heat rises, well sadly they've been lying to you, the heat doesn't rise by itself with no help, but it's the hot air raising which is cool, actually hot gases tend to rise and cold gases sink down.

How many earths do fit in the humongous sun? We could fit 1.3 million earths that's a proof that the sun is literally big.

The suns light only takes 8 minutes and 20 seconds to reach the earth, but it takes the light to get out of the core to the surface thousands and millions of years to give us some

warmth, so whenever you go out and feel the warmth of the sun be grateful.

What's the sun's expected life?

The sun is actually in the middle of its life, but in 5 billion years the sun is expected to go red giant, and when it becomes a red giant it will eat every single planet till Earth so mercury, Venus and the earth will say goodbye, but when the sun becomes a red giant after 10 billion years it will become a white dwarf looking like a beauty in our solar system, and in 10 quadrillion years it will become a black dwarf drifting in space all alone.

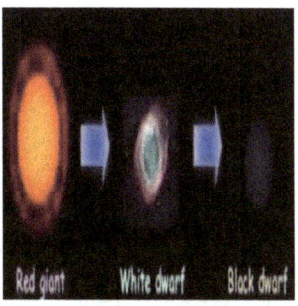

The Sun's Magnetic Field flips every 11 years, which causes a lot of problems to the earth's technology. Like in 2025 the suns magnetic field will flip, which means that the sun will shoot hot lava blast at the earth.

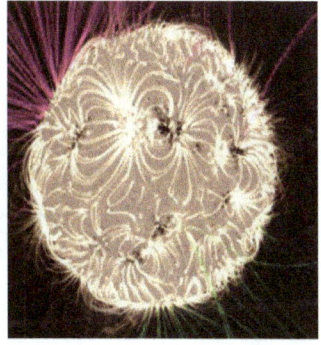

But how did we know everything about the sun? Well scientists can tell everything about a star by just looking at its color.

Photosphere

Photosphere is the surface of the sun. It means "sphere of light" the glowing thing you see in the day sky is the photosphere; the surface of the sun is the only thing that we can see from the earth 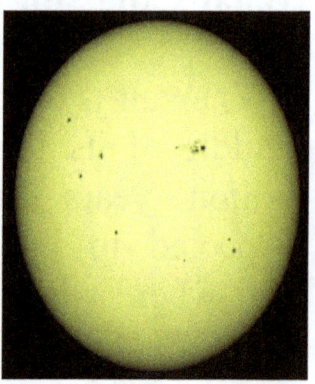 on a normal day, without any specialized equipment of course. Here, the temperature is about 6000 Kelvin. It is the thinnest layers of the Sun with thickness of 500 km. it also called the optical layer because the light goes out from it to outer space.

Chromosphere

A thin layer of plasma that lies between the Suns visible surface (the photosphere that we talked about before) and the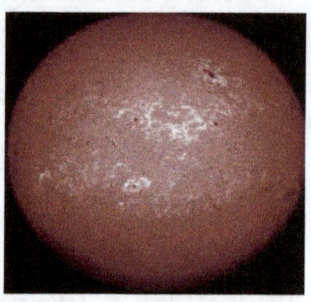

corona (we still didn't talk about) is the chromosphere. It at least extends 2000 km above the sun's surface. This layer can be seen as a red beam only when the total eclipse occurs. The temperature is about 10000 Kelvin.

Corona

The corona is my favorite layer of the sun. It's the outermost part of the sun; the corona can be seen during a total solar eclipse (only), it's usually hidden by  the beautiful sun's bright light, which makes it difficult to see it until as I said a full solar eclipse, in this region the temperature is about 2 million Kelvin, the reason for this high temperature is due to the huge energy that the magnetic field pumps into the Sun. Also, we cannot see this region from the ground all the time because its transparency and its warm gaseous nature.

It is full of meteoric nozzles and cracks in its crust.
Its color tends to yellow with white and gray spots.

Transition Region

This rejoin is very narrow (100km) layer between the chromosphere and the corona. This rejoin is known for its high temperatures it rises there abruptly from 8000 to a little close to 500,000 K!

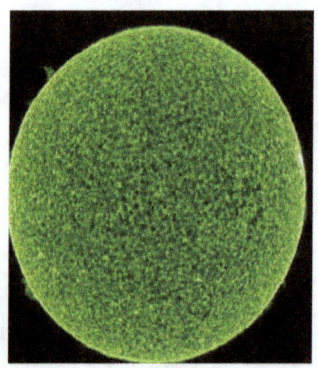

Mercury

Mercury is the closest planet to the sun, it's like the suns baby, a day on Mercury is really hot and in the night it's really cold. **Fun fact about mercury:** it's little bigger than the moon.

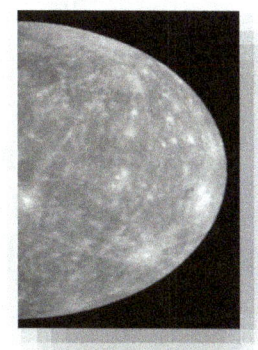

Mercury looks like the moon when you point a telescope on it, that's because it doesn't have an atmosphere so asteroids, comets, etc.… hit it making it look like the moon.

How many moons does mercury have?
Mercury doesn't have a moon, but why? Mercury is so close to the sun that it makes it impossible to have any moons, the sun's gravitational force might break

mercury's moon, but why would this happen? It would happen because the moons gravity would sadly cause tidal effects on mercury.

When can we see mercury in the sky with our naked eye?

1- Before the sun rises in the morning, we can see mercury.

2- After the sun sets, we can also see mercury.

How many spacecraft's went to mercury? Two, yes two, but which ones were they? It's actually Mariner 10 and MESSENGER that was launched by the USA.

Marnier 10

Mercury likes everything to be simple?

Yes, it is even simpler that cutting a paper with scissors. It doesn't have any moons; it doesn't have an atmosphere and it's covered with craters.

Does Mercury support life?

That's a question that needs to be answered. First, Mercury does not have an atmosphere that can save us from asteroids, comets, and other dangerous things. Second, Mercury is not making up its mind; mercury in the day is like touching the sun with your bare hands and in the night Antarctica but 30x the negative degree so we will burn to death in the day and in the night if we survived the day we would instantly freeze in place. In conclusion Mercury doesn't support life at all.

How long are a year and a day on Mercury?

A day on mercury is long it's 58.6 earth days long so if you feel a day here on earth is long then enjoy your day on mercury. A year is much longer than a day its pretty much 88 earth days so three day on mercury is two year so a year there is only 1.5 Mercury days.

How big is Mercury?

Mercury as I said is a little bigger than the moon, but its size is mercury is 4,879 km in diameter, which is only a little over a third in earth's diameter it's so small, so it's a little hard to see mercury without any binoculars or telescopes.

How strong is mercury?

If you suddenly fell on mercury, it's going to take a lot of time to get to the surface, because mercury will pull on you only a third of earth's gravity.

What is little Mercury made of?

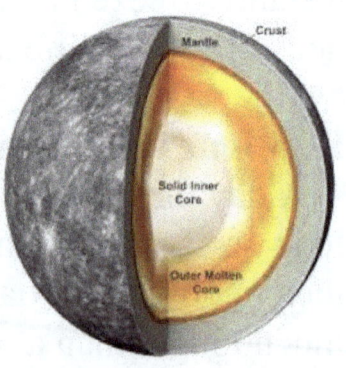

Mercury has an iron core. So, in perspective mercury has the largest iron portion in the solar system so whenever a planet needs iron it will go to mercury I guess, the outer layer of our loved mercury is made of silicates.

Mercury Facts:
It was known since ancient times because it can be seen easily.

Mercury had two guests: Mariner 10 and MESSENGER.

It orbits the fastest around the sun.

Mercury does look cool and cold but don't let it fool you mercury's surface temperature can come from -180 Celsius or 430 Celsius.

Venus

Venus is the second closest to the sun, and it's called the earth's twin, that's because Venus and the Earth are so similar in size. Venus is the hottest planet in the solar system, because it has a really thick atmosphere, so when the suns warmth comes to say hi to Venus it's locked there forever, but on the clouds it's habitable for a human, by containing oxygen and 95% the earth's gravity.

When can we see Venus with our naked eye?
We can see Venus, before the sun rises if it's ahead of the sun,
but when it's behind the sun we can see it in the evening.

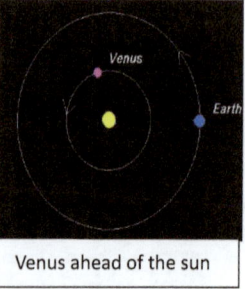
Venus ahead of the sun

NASA has some big plans for Venus, since on Venus's clouds there is oxygen and 95% earth's gravity and NASA's plan are to make cities on its clouds.

Have you ever wondered why Venus has no moon? Well Venus is so close to the sun, so it has the same problem as mercury.

Did you know that a day on Venus is longer than a year?

A day on Venus is actually 243 earth days long and the year is 225 earth days, but why? That's because Venus takes 243 earth days to orbit around itself and it takes 225 earth days to orbit around the sun.

The longest spacecraft that has ever survived on Venus was Venera13 in 1981 and it lasted only a little before 2 hours.

How big is Venus actually?

Venus has the diameter of 12,100km or you could say 95% the size of earth. The size similarity is only one of the reasons they're called twins another reason is there is found similar minerals on Venus that we have on earth.

Is there life on Venus?

This is a question that doesn't have an answer to yet so let's talk about what we discovered. First, Venus has such a thick atmosphere and it's too close to the sun so it's really hot on Venus hotter than mercury, because temperatures on Venus can get really high it can get to 464 Celsius on normal days so how would it be on hot days we would melt in Venus's atmosphere. Second, when Venera 13 was on its mission it saw that Venus has clouds that may support life as we know it by containing 95% of the gravity here on earth and some oxygen. So what is the truth does Venus does it support life or not?

What is Venus made of?

Venus as we said it's the earth's sister, **but what is it made of?** It's like the earth it has a crust a mantle made of silicates and iron in the core, there are similar things on Venus here on Earth.

How strong is Venus?

Venus is a planet that is like earth so does that mean they could potentially have the same gravity force? The answer is yes but it's not the same, if you suddenly fell on Venus, it would pull nine tenths here on earth, so they do clearly love each other.

What is the future for exploring Venus?

So, Venus is a planet that we still don't know a lot about so what are our plans for the future? Venus is going to have a lot of guests because it keeps melting everyone,

but we always want to explore Venus, because it's the closest planet to us, and it also reveals many and many clues on how the earth is working (Earth is so complicated). There are lots and lots of proposals to get space craft's going to Venus and then they will deploy balloon that might allow scientific instruments to travel freely in the atmosphere. Actually, we have sent already sent 2 of this idea to Venus; they are called Vega 1 and Vega 2.

Venus Facts:
Venus has an active surface including volcanos.

Venus spins the opposite side as the earth and other planets.

Venus was known from ancient times because it can be seen easily, and it was called the god of love if they knew that it was literally like hell, they would've called it the god of war.

Venus has been kicking its guests like: Mariner 2, Mariner 5, and Mariner 10 by melting them in the atmosphere and the longest one that survived was Venera 13.

Venus is a similar planet to earth.

Venus is a planet that hates spacecraft's it melts every time.

Its atmosphere is very dense and consists of carbon dioxide, water vapor and little oxygen, so occur greenhouse effect.

Earth

Earth is the 3rd closest planet to the sun and the only habitable planet in our entire solar system. The earth is a rocky planet with continents and oceans. The earth's atmosphere makes the temperatures good for us humans.

We can't forget that the earth is the biggest planet between the terrestrial planets bigger than Mercury, Venus, and Mars.

What are its moons like? The earth has one permanent moon called Luna, or mostly known as the Moon.

The Moon is about 240,000 km away you can fit every planet in the solar system between us and the Moon and still there would be 2000 kilometers left.

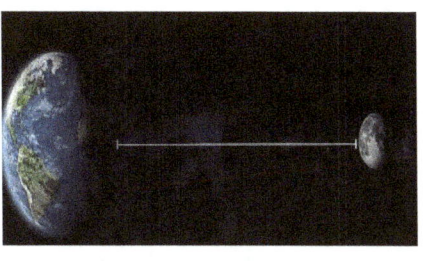

The layers of the earth are:
1- Inner core
2- Outer core
3- Mantle
4- Crust

How old is the earth?
　　The earth is around 4.6 billion years old.

　　The earth has so many oceans that water covers 70% of the earth's surface.

What is our atmosphere made of?

The earth's atmosphere is mostly made of nitrogen and it has a lot of oxygen for us to breathe non-stop, let's not forget to mention that our atmosphere protects us from meteoroids, asteroids, comets, etc. most meteoroids and asteroids break up before they could strike us.

Did you know every year our days become longer?

Every single year the moon moves 4-inchs from the Earth which makes our days a bit longer, but in 1 billion years the days on earth will be 25.5 hours.

Did you know that we were known since ancient times?

I know you're going to say of course we live here, but we didn't know our place in the solar system for such a long time, ancient people used to think that we were in the center and the other planets were orbiting us, which is ok, because they literally didn't have modern technology like now days.

Is there life on Earth?

Well the answer is obvious it's yes of course but what do other planets think? Well the Earth has everything needed for humans it has water which is really important, it rains here water not like Uranus and Neptune because in there it rains diamonds, and it also has plants which are really good because there is  some plants that we get food from like eggplants. Let's say you were born on Mars you might see us a bit different than what you look like, because you would be red and yes red, I mean that's kind of obvious since Mars is the red planet.

What is the Earth's surface like?

The earth is one of the few rocky planets in our solar system, so its surface is solid with lots of amazing Valleys, Canyons and beautiful Mountains, as we all know the earth is almost an ocean world with water covering 70% of Earth's surface (I'm telling you a little secret don't let anyone know there is an ocean world that is a super earth its name is Kepler-138 d this exo-planet is covered with water all over it's like the planet in interstellar, by the way interstellar is a movie).

How was the earth formed?

Hmmm… the earth is our favorite, **but how was it formed?** The earth was formed like all the terrestrial planets it was formed when the gravity pulled swirling gas and dust, so it could become like how it is now; did you know that the Earth 4 billion years ago was never habitable?

What would happen if a black hole appeared here on earth?

So if you don't know what black holes are, basically black holes are the strongest objects in the universe they are stronger than big stars like be Betelgeuse which is the biggest star in the universe anyways they could be from stellar to ultra-massive also if a black hole doesn't have a lot of mass it doesn't mean that they could swallow everything, one last fact about black holes is that they might be the last object in the universe. Now let's get to what we are here for, if a black hole suddenly appeared on earth we wouldn't be affected a lot if it didn't have a lot of mass, but it would pass on us just giving minimal damage, but if the black hole was large oh boy we would have some serious trouble, because it will pull some things towards it and since black holes eat light faster than blinking it would be pitch black on earth, but to note that the chances of a black hole appearing are slim so we might not have to worry a lot.

So, countries here on earth are fighting a lot (sadly) so what if a country decided to bomb us with a nuclear weapon what would happen?

So, the chances of a country bombing us are slim, because there would be some serious damage, because if a country bombed us, we would lose a lot of the population, because a lot of people would die and the earth itself would be damaged, because we might have a nuclear winter buildings would be damaged  severely. So that's why all countries are trying not to make any wars let's not talk about the wars that now are happening but now the countries that are in war don't make a lot of trouble so they wouldn't have to use the nuclear weapon.

What if the earth was hit by a needle at the speed of light?

So if a needle was moving at the speed of light through our atmosphere, it would have the equivalent kinetic energy of 100 tons falling of a 15-story building but if it hit the earth's crust we would see the biggest explosion ever, the people living near the explosion would die immediately, but people further away would be burned by the heat waves, but if the needle wants to go even further the earth would turn into a giant fire ball, this might never happen cause nothing can travel at the speed of light.

Is the earth flat or a sphere?

If the earth was flat the oceans would meet up together in the middle and the suns light would only come in the middle not anywhere else and also the moon would have a line in the middle of it, so the earth is not flat.

Earth

Its mass is about 5.972 × 1024 kg, the distance between earth and sun is 150 million meters. which is known as 1 astronomical unit (AU).

Its surface temperature varies from minus 89.2 to 56.7 Celsius degrees.

It consists of:

1. Crust, which includes the outer solid surface of the earth and the bottom of sea and oceans, and it's about 5 km in depth under the bottom of oceans and under the solid surface about 48 km in depth.

2. Mantle, which extent to 2880 km under the earth surface, consists of several minerals of silicates, magnesium and iron.

3. Core:

a. outer core; its depth range between 2900 to 5100 km, which consists of liquid nickel and iron, and due to these rotations inside the earth, as a result the magnetic field creation become a fact.

b. inner core; is also from nickel and iron but in solid structure. It has a spherical shape with a radius of 1280 km.

Earth Atmosphere surrounds the earth surface from all directions, where the sea surface level represents the lower limit. It consists of 78% Nitrogen, 21% Oxygen and 1% the rest from carbon dioxide, vapor, argon and ozone. (the major reason for nitrogen is to reduce the oxidation of oxygen). The density of gases in earth atmosphere decreases as we move far away from the surface.

Atmospheric sections:

1. Troposphere; the region from the earth surface up to 11 km. Here is the weather variation space, as we go 1 km upward the temperature decease to 7 Celsius degrees.

2. Stratosphere; the region from 11 km up to 50 km. The ozone layer is between the troposphere and this region, which is 20 km in thichness.

3. Mesosphere; the region from 50 km up to 85 km. Here the temperature decreases rapidly until it reaches minus 90 Celsius degrees.

4. Thermosphere (Ionosphere); the region from 85 km to 700 km. Which contains charged particles that come from sun radiation (ultraviolet and infrared) that ionize the gas atoms of the atmosphere. As a result, this region can reflect radio waves.
5. Exosphere; here it extends from 700 km up to 35000 km.

There is no magnetic field on moon because there is no molten core and the rotational velocity around itself is low.

Its surface covers the conical volcanic craters formed by the meteorites collision with the moon surface. It has no gas atmosphere or clouds or winds, also it looks black.

Solar Eclipse occurs when the moon blocks the sun light to viewers on earth .Some or all the sun seems to disappear. That happens when the moon lies directly between earth and sun.

Lunar Eclipse occurs when earth blocks out the sun light shining on the full moon, making the moon disappear fully or partially. This happens when earth lies directly between the moon and sun.

Moon

The moon is the only natural satellite for the Earth. It's also loved by astronauts, because we sent a lot of rockets with astronauts there, we discovered a lot of things and we discovered its formation, **so how was the moon formed?** The moon was formed by a planet that was the size of Mars, so this planet hit the Earth and tiny rocks were flying, but they all  formed the moon that we know today. Actually, there are 12 people that landed on the moon and they brought back objects from the moon that were studied here on Earth. **Did you know that the last person that step foot on the moon had an allergic reaction?** The astronaut Harrison H. Schmitt during the Apollo 17 Mission in 1947 had an allergic reaction, **why**

though? He discovered that he had severe allergic reaction to moon dust, so he is allergic to the moon. He had a running nose and itchy eyes when he was exposed to moon dust.

How many phrases does the moon have?

The moon has 8 phrases called: New moon, Full moon, First Quarter, Last Quarter, Waxing Crescent, Waxing Gibbous, Waning Gibbous and Waning Crescent.

How does solar eclipse happen? Solar eclipse happens when the full moon is right

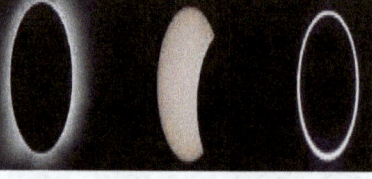

in front of the sun but the sun, earth and the moon must be in a straight line every month there is a full moon, but not every month there is solar eclipse, because the moon is either 5 degrees up or 5 degrees down.

How long are a day and a year on the moon?
The day on the moon is 29 Earth days, and a year there is 354 Earth days, in perspective it takes the moon 354 days to orbit around the Earth.

Moon Facts:
The moon is the only place in the universe were human step foot.

The moon is the brightest object in our sky, and the second is Venus.

The moon is the 5th largest moon in the solar system between 200+ moons.

The moon is nearly twice as big as Pluto and 2 times smaller than mars.

We can see 59% from the moon.

Do you know why the moon is called the moon simply?

It's because people didn't know there is other moons until Galileo Galilei discovered four moons orbiting Jupiter in 1610.

What would happen if the moon came a little near earth?

If the moon were to come a little closer to Earth, several significant changes would occur. The increased proximity would lead to intensified tidal forces, resulting in more frequent and severe coastal flooding. Earth's rotation would be affected, with slower days and potential disruptions to weather patterns and ecosystems. The closer moon could also destabilize Earth's axial tilt, causing unpredictable climate variations and irregular seasons. Geological activity would increase, with more earthquakes and volcanic eruptions. Additionally, while space exploration would benefit from easier lunar missions, there would be navigational complexities and higher collision risks. While this scenario is unlikely, it highlights the delicate balance and interconnectedness of celestial bodies in our solar system.

What is on the other side of the moon?

The moon that we see every day is only showing us one side of it, **so what's on the other?** The other side of the moon has simple things like craters that cover it, but these craters are less smooth dark spots unlike the other side. (Maria is a crater that covers the side shown to earth), we never knew what was on the other side of the moon until we had advanced technology that can travel to space.

What if the moon suddenly disappeared?

We would be absolutely doomed, because if the moon left us all alone we would either have no seasons or the worst option which is extreme weather or even ice ages could you imagine that. Also, let's not forget to mention that the moon also has a strong effect on the earth's gravitational force, so let's not hope that the moon would leave us all alone.

Radius	diameter	gravity	Density	Mass	Ring system
1.737.4 km	3,474.8 km	1.62	3.34	7.34767309 × 1022 kg	No

Mars

Mars is the fourth closest planet to the sun and the last planet from the terrestrial planets. Did you know that even though mars is smaller than the earth yet it has 2 beautiful moons called Phobos and Deimos, but our moon is bigger than these two moons combined?

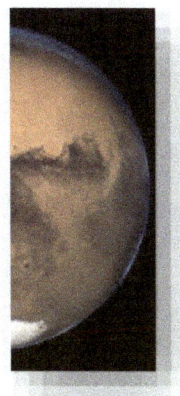

Do you know why we're called the terrestrial planets (Mercury, Venus, Earth, and Mars)?
That's because we're the only planets in the solar system with solid surface, we're like rocky planets and we have no rings like in the outer planets (Jupiter, Saturn, Uranus, and Neptune).

Fun fact about Mars: Mars is actually 2× smaller than the Earth and 2× bigger than the Moon.

Have you ever wondered why Mars is actually called the red planet? Mars is considered the red planet, because of its rusty surface and mars is covered with Iron rust.

This is Mars surface which was taken by Mars's rover.

Did you know this?

Mars, Venus and the Earth in the begging of the solar system used to orbit in the same zone, which was the habitable zone.

What are our dreams for Mars? Some of our dreams for Mars is to make it habitable and now Elon Musk is going to send bombs into mars so it would have holes so the sun light can heat up Mars, he also said that the first humans on Mars will be in 2033 or 2035,

but the first 1mil people on Mars will land in 2050.

Did you know that the largest Volcano in the solar system is on Mars?

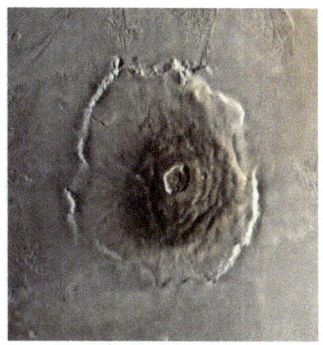

Well meat Olympus one of Mars's volcanos and it's the largest volcano in the solar system its 3× taller than Mount Everest, if you want to climb this volcano it would literally take 7 months without stopping you might want to hold your breath, because Mars atmosphere is so thin.

What is Mars's surface like?

Mars has a beautiful surface filled with lots of canyons, Volcanos not like earth (Earth has volcanos but not a lot compared to mars), dry lake beds and last on this list is craters, but it's mostly covered with red dust, but mars also has clouds like earth so it's not only earth, but sometimes these clouds blows the red dust to turn into a dust storm I feel bad for Mars surface because it must deal with lots of dust storms.

What is Mars's atmosphere made of?

Now Mars used to be habitable in the past, but it seems like all the oxygen went away when it became inhabitable, because 95% of Mar's atmosphere is made of Carbon dioxide, but what is the 5% that is left made of? Well 2.6% which is a bit over half is made of molecular nitrogen, so now we have 2.4% left let's see what it is made of well 1.9% is made of argon and 0.16% is made of molecular oxygen and last on the list is carbon monoxide with only 0.06%

What were the missions to mars in the past?

Mars had lots of guests in the past, mars had a total of 50 guests in the past starting with Mariner 9 which was launched back 1971 it ensured victory since it was the first space craft that ever landed on a another planet, second one on our list is NASA's Viking Project it was the first successful mission on mars and it came back with pictures of the surface later studied, but these were 2 of the most successful missions on Mars.

Is there life on Mars?

Well mars may or may not support life here let me tell you. Mars as we said has a thin atmosphere so we will not be protected as much like here on earth, because mars is kind of close to the asteroid belt, so there might be a lot of asteroids going for mars to hit it and we will not have enough time to move from where we are and mars is mostly made of carbon so we will not survive on mars, because we need lots of oxygen, but mars doesn't have a lot of oxygen let's not forget also that mars is far from the sun so it means that temperatures will not be very good for us because we would most likely only survive 90 seconds, so mars does not support life I don't know how we will have 1 million people on mars by 2050?

What is the future of exploring mars?
  Our little planet mars had been explored since the 20th century, we discovered lots of things on mars like the ancient ocean that are frozen, we always looked for signs of potential life on mars or even past life, but we discovered that mars was habitable 4 billion years ago, we had some of our missions fail, but we still went back for mars, but what is the future of exploring it? First, we plan on sending lots of spacecraft's to explore mars and India is planning to launch Mars Orbiter Mission 2 in 2024 and we have human plans to mars as we know the 2050 mission by Elon Musk.

Mars Facts:
Mars has such a thin atmosphere.

Mars's atmosphere is active, but Mars's surface isn't active it's sleepy, which means no volcanoes are alive, they're dead.

Mars was known since ancient times because it can be seen easily, and they used to call it God of war if they knew how sweet mars is they would've called it God of love.

Its atmosphere is very close to Earth, also its surface contains volcano deposits.

It has two small moons: Phobos and Deimos.

Mars/Phobos

Phobos is Mars's biggest moon it was created by tidal forces; Phobos was almost shattered by a giant crater and beaten by thousands and thousands of meteorites, on a collision course with Mars, let's not forget that Phobos is so close to Mars that some parts of Mars can't be seen because of Phobos and it orbits Mars 3 times a day! Phobos is getting closer to Mars every year it comes 1.8 meters closer and probably in the next let's say 50 million years it will either crash into the surface of Mars or get broken up giving Mars rings so it will be small Saturn but in terrestrial planets.

Is Phobos that big in its diameter?

Phobos may seem big like what we saw in the picture, but it's not that big even it's only 22.533 km in diameter it's not that big at all.

How old is Phobos?

Phobos is around 4.5 billion years old kind of like the earth at this point.

What is Phobos made from?

According to studies Phobos is a type C-rock which is similar to blackish carbonaceous chondrite asteroids.

Does Phobos have water on its surface?

Phobos does have water, but the water on Phobos is less than any object in the solar system.

Mars/Deimos

Deimos is the second moon of Mars, it's like its big brother Phobos it's small and chubby and of course an object filled with craters, but Deimos's craters are smaller than Phobos's a bit. It was discovered on Aug. 17, 1877 by Asaph hall. Also, it's made of a type C-rock like its brother Phobos. Deimos takes 30 hours to complete one full orbit around mars just a little over a martin day! Fun fact about Deimos: it looks like a potato.

Deimos Facts:
   It doesn't have an atmosphere like Phobos.

Phobos and Deimos are closer to type C-rock asteroid than a moon.

All the other moons are spherical except Phobos and Deimos.

There haven't been any specific missions to Phobos and Deimos I guess we hate the both of them.

Asteroid Belt

The asteroid belt is between Mars and Jupiter and it's literally huge and there are some big asteroids over there and they're beautiful like

That is 4 Vesta and it's the second largest asteroid in the asteroid belt and its 525km across.

this is Ceres the biggest asteroid in the asteroid belt and it's 939km across.

Do you know what is the asteroid belt made of?

It's actually made of asteroids, ice and comets.

What are most asteroids made from?

Almost (3 out of 4) asteroids are made of Carbon- based rock.

How was the asteroid belt formed?

About 4.6 billion years ago, the solar system was a disk of gas and dust known as the solar nebula. Within this disk, the young Sun began to form at the center, while smaller clumps of material called planetesimals formed in the surrounding regions. These planetesimals were the building blocks of planets and asteroids. The gravitational influence of Jupiter, the largest planet in our solar system, played a crucial role in shaping the asteroid belt. Jupiter's powerful gravity caused disturbances and gravitational interactions with the planetesimals in the region between Mars and Jupiter. These interactions prevented the planetesimals from accreting into a larger planet, disrupting their growth and scattering them into various orbits. Over time, collisions and gravitational interactions among these planetesimals led to the formation of the asteroid belt. However, it's important to note that the asteroid belt is not a dense collection of asteroids as often depicted in movies or illustrations. The asteroids are spread out over a large area, and the total mass of all the asteroids in the belt is estimated to be less than the mass of Earth's Moon. Most asteroids in the belt are relatively small, ranging in size from a few meters to hundreds of kilometers in diameter. In summary, the asteroid belt formed through a combination of gravitational interactions and disruptive forces from Jupiter during the early stages of the solar system's formation.

Is the asteroid belt that important?

The asteroid belt may not seem that important, but it's really important it saves us from a little snake called Jupiter, why did I call it a snake? Because Jupiter and its fellow friends (Saturn, Uranus, and Neptune) are all gas giants which means they have strong gravity, so they are going to try and eat us, well you are going to say that there is nothing between us and the sun, so why are we not in danger? Well, the sun is pretty far from us so we're mostly protected. Jupiter saves us from the asteroids it pulls them towards it so this is why the asteroid belt is important!

Asteroid Belt Facts:

The asteroid belt protects us.

The asteroid belt was formed 4.6 billion years ago.

The asteroid belt has between 1.5 million asteroids to 1.9 million asteroids.

16 psyche (one of the biggest asteroids) is estimated to be worth $10,000 quadrillion dollars which is more than all of the money here on earth since on earth there is 108 trillion dollars only compared to how much its worth.

Ceres

Ceres is the biggest asteroid in the asteroid belt, but Ceres was too big for its neighboring friends so in 2006 astronomers confirmed that Ceres is now considered a dwarf planet. Even if Ceres does not have a moon or an atmosphere it still has something not lots of planets 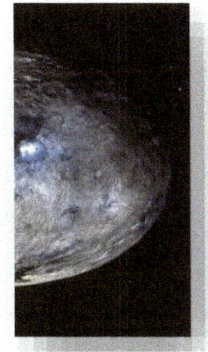 have, which is water yes you heard me water on Ceres. Ceres has almost third of the asteroid belts mass, but still it doesn't mean it's very big because Ceres is still far from being the size of our moon. Ceres has lots of Ice underground and it's heavily covered with craters.

Radius	476 km
Ring system	No
Moons	None
Mass	939,300 kg
Diameter	946 km
Planet type	Dwarf
Gravity	0.27 m/sec

What is little Ceres made of?

Ceres is mostly like the terrestrial planets. First, Ceres has a layered interior like the terrestrial planets, even though Ceres layers are not defined clearly. Second, Ceres has a pretty normal solid core, but its mantle is much cooler than the core since the mantle is made of ice water how crazy is that?! If it's true that Ceres is composed of 25% of water then Ceres would have more than here on Earth! Ceres crust is interesting, because Ceres crust is made of rocky and dusty salt deposits, but Ceres salt is not what you think, Ceres salt is made of different minerals like magnesium sulfate.

What is Ceres surface like?

Ceres is covered with tiny small, young craters and they are countless but none of the craters there is more than 280 kilometers in diameter. This is sad since this little planet lived for 4.5 billion years it must've been hit by some lots of asteroids cause honestly it is living in a risky zone.

What is Ceres atmosphere like?

Ceres has an atmosphere and it's really thin thinner than Mars's atmosphere, but astronomers say that there is water vapor in there and it's been proofed. The vapor might've been produced by ice volcanos which is cool or by the ice near the surface sublimating of course turning from ice into solid.

Does Ceres have a magnetic field?

Ceres is not thought by astronomers to have a magnetic field.

Jupiter

Jupiter is the 5th closest planet to the sun and it's the biggest of all the planets. Jupiter is literally so big that you can fit 1,300 Earths inside of it. Also, Jupiter is our warrior it saves us from all asteroids and comets it takes to itself. Jupiter's magnetic field is biggest thing single thing in our solar system it is 26 million kilometers which is close to 20x the size of the sun it extends to past Saturn's orbit! If we wanted to see it in our night sky it would be five times bigger than the moon.

Does Jupiter have rings? Yes, it does. Its rings are so dark that we didn't see them until a spacecraft made it there it's like Neptune and

they're made of tiny particles that meteors kicked.

Did you know that the great red spot- on Jupiter is 3.5% bigger than the Earth?

We might lose the great red spot in the next 20-30 years, and then the great red spot is going to leave us alone. So, we talked about the great red spot but what happens in there? There happens waves that go up to 600km/h and the waves have been going for 300 years!

How many moons does Jupiter have?

Jupiter in 2019 had 13 moons, but Jupiter's gravitational force made small asteroids and whatsoever become its moons. Now Jupiter has 95 moons, so it's the planet with the 2nd most moons, but maybe in 2025 if we recheck Jupiter's moons they'd be 200?

Why do gas giants get to have so moons?
Gas giants get to have these many moons, because they have such a big gravitational force. Also their humongous sizes. Jupiter even has the biggest moon in the solar system: Ganymede.

What is Jupiter's surface like?
There is no surface on Jupiter because Jupiter is a gas giant, let's say you managed to get inside Jupiter you're going to fly non-stop, why did I say you managed? Jupiter has a strong gravitational force as I said, so you will get ripped apart. This planet only swirls water and gas so let's say we launched a spacecraft to Jupiter well my friends Jupiter has no surface to land, you'd say the spacecraft could fly through Jupiter; well, it is going to get absolutely destroyed before even reaching Jupiter due to Jupiter's magnetic field.

What is Jupiter made of?

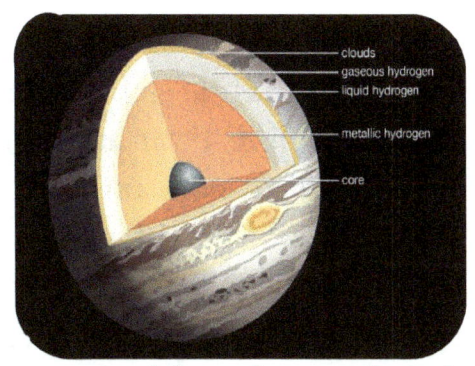

Alright, we all know that Jupiter is the biggest planet, but what is the biggest planet made of and is there something that makes it special? Well, Jupiter our fellow friend is mostly made of hydrogen and helium just like the Sun; deep in Jupiter's atmosphere spicy things happen like: Pressure, temperature increase and compressing the hydrogen gas into liquid, but what does that give us?! It gives Jupiter the nickname of: The planet with the biggest ocean. Jupiter's ocean is not just any ocean its special, since Jupiter's ocean is made of hydrogen not water!

How strong is Jupiter really?

Now size doesn't matter we all know that, why did I say that? Well, my friends if Jupiter is big it doesn't mean it has strong gravity, since Jupiter is literally made of gas, it made its gravity weaker than what we expected, since if you were falling on Jupiter it will just be 2.6 times stronger than here on Earth, let's say you weight 45 kg here on Earth, but on Jupiter you are going fat my friend since on Jupiter you would weight 108 kg oh boy hope we don't go to Jupiter at any time (because when you enter Jupiter you would be killed instantly since Jupiter has toxic air!).

Jupiter Facts:

It's the planet with the 2nd most moons.

It's the largest planet.

It's made with the same things as the sun.

Jupiter needs to be 80× the mass of itself to become just a red dwarf star, but if it wants to be like our sun it would need to be 1,000 times its current mass.

The temperature on its surface is about minus 130 Celsius degrees.

It consists of a thick core surrounded by a layer of liquid mineral hydrogen and outer layer consisting of a large amount of molecular hydrogen.

The rotational period around itself is 9.925 hours and rotates around the Sun at 11years and 315 days.

It has many moons four of them are big in size called Galilian moons (Callisto, Europa, Io, and Ganymede)

Jupiter/Io

Io might seem a normal world but it's not. Io is the most active object in our solar system, since Io has a lot of volcanos interrupting they could go really high, a normal day there could be just volcanos interrupting. It's just a bit bigger than our moon, but it's the third biggest moon between Jupiter's moons and it's the fifth from Jupiter's orbit.

What is it made of?

Alright Io is such a beautiful moon which is made of Sulfur dioxide; Sulfur dioxide is a primary constituent of a thin atmosphere on Io. Io has no water which is sad.

What is Io's surface like?

So, as we said Io's is the most active volcano surface in the entire solar system, but what is Io's surface like? Temperatures on Io are low it's -130 Celsius there or very high till 1,649 Celsius and scientists' like to call Io the planet of "ice and fire" pretty cool name there. On the outside of Io and on the inside there is iron, as iron is on its core, and an outer layer is made of brown silicate.

What is the distance from Jupiter to Io?

The distance from Io to Jupiter is 2x the distance from the earth to the moon, since the distance from Io to Jupiter is 422,000 km.

Does Io have a magnetic field?

Io doesn't have a magnetic field on its own, but its father Jupiter has a very strong magnetic field, but it's thought from the Galileo space craft, that an iron core form Io's center may be is giving Io its own magnetic field.

When was Io discovered and by who was it discovered?

Io was discovered in 1610 by Galileo and it played a significant rule in the 17th and the 18th century in the development of astronomy.

Mass	Radius	Diameter	Ring system	Moons	Gravity	Day	Year
$8.93*10^{22}$ kg	1,821.3 km	3,643.2 km	No	None	1.796 m/s2	1.8 earth days	42 hours

Is life possible on Io?

Life on Io might be impossible due to its very scary temperatures since it could be very cold to unbelievably hot and also its crazy volcanos eruption which might kill all the people on there.

Io Facts:

Io is one of the few moons to have its own atmosphere.

Io has such a thin atmosphere thinner than Mars's atmosphere.

At any time someone is talking about Io there might be 9 volcanos erupting which is crazy.

Io is so close to Jupiter that there are tides on Io's surface.

Io is about 4.5 billion years old which is close to Jupiter.

Jupiter/Ganymede

Ganymede is the biggest moon in the solar system, it's the biggest moon without an atmosphere yet it has its own magnetic field, which is the only moon in the solar system with its own magnetic field. It's bigger than the two small worlds Mercury and Pluto. NASA Hubble's space telescope found great evidence about a saltwater ocean, which kind of makes it habitable for humans its ocean is 10x deeper than oceans here on Earth. Ganymede's Ocean is buried under 150 km crust mostly made of ice.

What is Ganymede made of?

Ganymede is an interesting moon. It is made of three main things: a spherical looking shell of

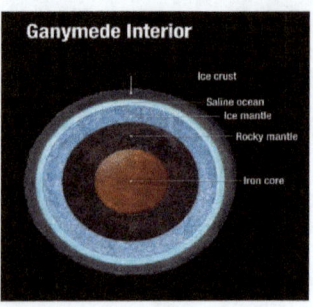

mostly ice that is surrounding the rock shell, a core made of metallic iron at Ganymede's center, lastly it is made of a spherical looking shell of rock which is the mantle and the mantle is surrounding the core.

What is it surface like?

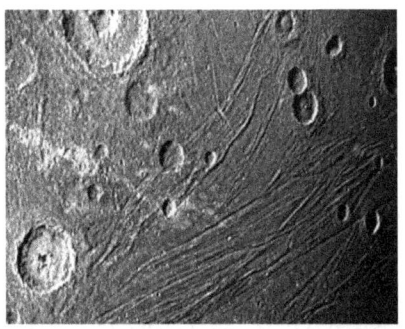

Ganymede had some space craft's visits to it and when they went there, they found out what is Ganymede's surface is like. So we are going to say what they saw, the first thing we must know about Ganymede's surface is that it is a mix of 2 types of terrain, Ganymede has 40% of its surface for cratered dark regions, I know Ganymede likes everything to be simple, but what is the other 60% made of? The other 60% is covered with a light grooved terrain, which forms something cool, which is an intricate pattern across Ganymede.

So Ganymede is the biggest moon but how big is it really?

Ganymede is the largest moon in the solar with the radius of 2,643.1 km which is 2.5 times smaller than Earth.

When was Ganymede discovered and by who was it discovered?

Ganymede was discovered by the same person that discovered Io, which is an Italian astronomer called Galileo Galilei back in the 7th of January in the year 1610.

Is life possible on Ganymede?

Ganymede might be possible for life I'm not lying, because like we said that Ganymede has an underground ocean and some scientists think that water and rock are the development for the possibility of life, so there might be life on Ganymede.

Does Ganymede have a magnetic field?

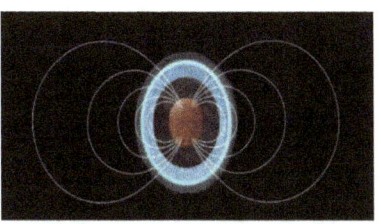

Ganymede does have a magnetic field, but how big is it? Ganymede's magnetic is 719 ± 2 Teslas, which is pretty big for a moon.

Ganymede Facts:
It's the third of Jupiter's moon.

The day over there is 7 earth days.

The year there is about 12 earth days.

Two giants, since Ganymede orbits Jupiter the biggest planet in the solar system and Ganymede is the biggest moon in the solar system, which makes Jupiter and Ganymede two giants.

It has a thin atmosphere (almost no atmosphere).

It is a popular destination.

Jupiter/Europa

Europa decades ago took everyone's mind; because people thought that Europa had alien life (it does have alien life which is bacteria). Europa is the 6th biggest moon in our solar system, which means it is smaller than our moon and it's the smallest between the four Galilean moons. Europa is an interesting moon that now we are going to talk about it.

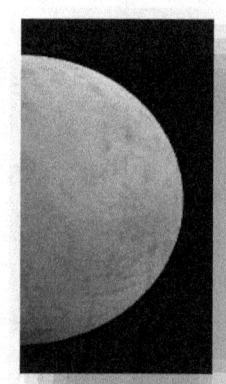

How big is Europa actually?
Europa is not really big it's 90% the size of our moon with the diameter of 3,100 km and the radius of 1560 km, this makes Europa the sixth-largest moon in the solar system, which is a great success.

Does Europa have an atmosphere and if it has what is it made of?

Europa has a very beautiful atmosphere that is only tenuounced atmosphere made of oxygen. In 2013, NASA announced that the Hubble space telescope found some evidence that Europa might be actively venting water into space, which is really cool. This means that this moon is geologically active in the present day which is amazing!

What is the distance from Europa to Jupiter?

The distance from Jupiter to Europa is pretty good, which is 671,000 km; it's almost triple the distance from the earth to the moon.

How long are a day and a year on Europa?

The year and the day there are pretty similar to each other, since a day there is 3.551 Earth days, but a year there is 3.551181041 Earth days so in a simplified way a day and a year there are 3.5 days.

What is Europa made of?

Europa is such a beautiful world made with lots of stuff and now we are going to say them, starting off with that Europa is an 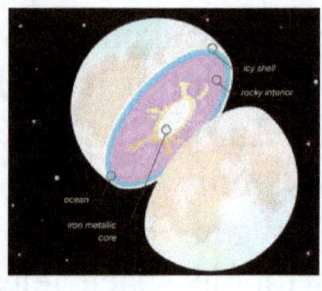 Icy world. Second, Europa's core is probably made of iron-nickel and Europa is mostly made of silicate rock, which is fascinating and we still didn't come to the fun part that Europa has a water-ice crust how crazy is that?! Finally, Europa's surface is made of frozen water, which is of course the smoothest in the solar system.

Is there life on Europa?

Sadly even though everything seems capable for life Jupiter is blasting Europa with radiation and Europa couldn't survive.

Europa Facts:
Europa is 4.5 billion years old!
It was discovered in 1610 like its fellow Galilean moons.

The closest that Europa could get to its father Jupiter is 664,861 km. This is called Perigee.

The furthest that Europa could run away from Jupiter is 676,936 km. This is called apogee.

Europa's atmosphere is so thin!

Europa might have as much as twice water than here on Earth.

Jupiter/Callisto

Callisto is the last moon from the Galilean moons, the Galilean moons are Io, Ganymede, Europa and Callisto. Callisto is the second largest moon of Jupiter  after Ganymede, but is the third largest in the solar system just after Ganymede and Titan it is as large as our little baby Mercury even though it is just only as third as massive. Fun Fact about Callisto: Callisto is the most heavily cratered object in the solar system!

How big is Callisto?
Callisto as we said in its introduction it is the third largest in the solar system. It has the diameter of over 4,800  km, which is amazing for a moon especially those moons are not known to be that big, but

let's compare it to Earth. Callisto is 2.6 times smaller than Earth, which is alright you could say.

What is the distance from Callisto to Jupiter?

Callisto is really far from Jupiter it is 1,883,000 km far from Jupiter! It is the furthest from the Galilean Moons!

How long are a day and a year on Callisto?

The day and a year on Callisto are the same they are twins! As of a day and a year there are 17 Earth days!! So, a century there is short. But why are a day and the year are the same?! Well, that's because it takes Callisto 17 days to orbit around itself and 17 days to orbit around Jupiter! The Galilean moons have special things!

What is Callisto's atmosphere like?

Callisto does have an atmosphere, but it is extremely thin and scientists call its atmosphere an exosphere, because of how thin it is! This was announced back in 1999 when the Galileo spacecraft saw an exosphere made of carbon dioxide now studies show that Callisto also has oxygen and hydrogen in its exosphere.

What is Callisto made of?

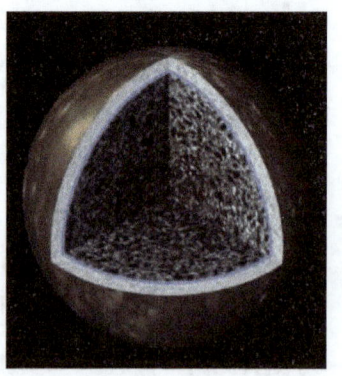

Callisto is an interesting moon, since it is the most heavily cratered object in the solar system and its surface is also filled with rock and ice, but what kind of rock is Callisto really made of? Well Callisto must really love rock since 60% of Callisto is made of rock/iron, but only 40% for ice it then it must not really like ice.

Is there life on Callisto?

Callisto as we said has oxygen, but does that make it habitable? Yes, it does if it had some serious amount of oxygen like here on Earth, but it kind of helped Callisto become one of the places that might support life beyond Earth in our solar system.

Callisto Facts:
There is no evidence (till now) that Callisto was tidally heated.

Callisto might've got its craters due to lots of asteroids and comets hitting its surface, its atmosphere is just too weak to protect its moon.

Callisto was once wanted from NASA as of NASA wanted to do basecamps for us humans.

Saturn

Saturn is the 6th closest to the sun and the most beautiful, it has loads of rings. Saturn is the most loved by children, adults and elders.
What are its moons like?
Saturn as we know is the planet with the most moons coming at 146 moons.

What are Saturn's rings made of?
There mostly made of asteroids, some pieces of comets and broken up moons that broke before even reaching the planet.

How long are a day and a year there?
A day on Saturn is short, since a day there is 14.7 hours and a year there is pretty long, since a year there is 29 Earth years.

What is Saturn's surface like?

Saturn is like the brother of Jupiter, so it's like Jupiter right? Yes, it is like Jupiter so I guess we don't have to explain again!

How big is Saturn?

Saturn is literally big it is the second largest planet in the solar system (just after Jupiter), with the diameter of 116,500 km, which is 10 times the size of Earth (in diameter)!

What is Saturn made of?

Jupiter and Saturn are literal twins, since Saturn is just like Jupiter with it being a gas giant and also Saturn has no surface with it swirling gases and liquids. Saturn's core is made of iron and nickel.

How strong is Saturn?

Saturn is strong, because the surface gravity is 107% of the surface here on Earth, so let's say you put a platform on Saturn (which is nearly impossible I'd say) let's say you weight 45 kg, on Saturn you would be fatter, because you would weight 47.84 kg so you'd be 2.89 kg fatter.

Saturn Facts:

Saturn and Jupiter are both gas giants, they're mostly made of: hydrogen and helium, just like the sun

Saturn has a really thick atmosphere.

Saturn has 7 main rings with spaces between each of them, it's such a beauty.

Saturn was known since ancient times because it was seen easily with its stunting rings.

It's the second biggest.

It's the planet with the most moons.

It has so many rings that are bright not like Jupiter's, but an exo-planet called J1407b also known as super Saturn has 600× Saturn's rings.

It has a very strong magnetic field which comes from the high electric currents of the Hydrogen minerals inside Saturn.

Saturn/Titan

Titan is the biggest moon of Saturn and the second biggest in the solar system, this moon is bigger than all the dwarf planets here in the solar system like: Ceres and Pluto. Titan is the only known object in the solar system and space other than Earth to have stable bodies. Titan is the only known moon to have a dense atmosphere.

How big is Titan?
Titan is literally big it is so much bigger than our moon with the diameter of 5,149.5 km. Titan is literally big as we saw if it was not one of Saturn's satellite it would've been considered a dwarf planet.

What is Titan's atmosphere made of?

Titan's atmosphere is similar to Earth's atmosphere but with a plot twist. Titan's atmosphere is mostly made of nitrogen (95%) like the Earth, but Earth is fewer than Titans percentage the plot twist is that it is also made of methane (5%) and it is also has some small amounts of Carbon-rich compounds. High in Titans atmosphere, methane and nitrogen molecules are spilt apart, because of the Sun's ultraviolet light (the sun is scary) and by high-energy particals accelerated in Saturn's magnetic field (I guess both Saturn and the Sun hate Titan).

What is the distance from Titan to Saturn?

Titan's is pretty far from Saturn with 1.2 million km now that's pretty far for a planet with its own moon, now imagine the moon was 1.2 million km away our days would be long!

What is Titan made of?

Titan is a cool moon it is icy, like Ganymede, its surface is amazing (lucky for the aliens if 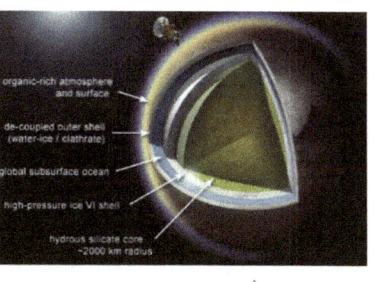 there is), because Titan's surface is rock-hard water ice, how cool is that?! We still didn't talk about this fact, which is both crazy and awesome Titan has water beneath its surface!

How strong is Titan?

Titan is big but not strong if you were to fall on Titan, you would fall at the speed of 1.353 m/s2 (14% of Earth's gravity), which is 7.3 times weaker than the Earth I hope we don't jump or fall on this moon because we will be in the air for such a long time.

Is life possible on Titan?

There is no evidence till now for life on Titan, but Titan has some suitable conditions for life as if there is an ocean on Titan, but there isn't enough oxygen for one person so you would probably survive a day on Titan.

Titan Facts:

Titan was discovered back in March, 25th 1655.

A day and a year are the same; the day and the year are 16 Earth days.

Titan is larger than Mercury.

Its surface is in deep freeze at minus 180 co.

Titan is the only moon in solar system to have a dense atmosphere which consists of methane.

Saturn/Mimas

Mimas is one Saturn's amazing moons, since it's the second most cratered moon in our solar system, Mimas is tidally locked. Mimas is like the moon we just see one side of it, Mimas likes to let Saturn see one side of it.

How big is Mimas?

Mimas has a decent size for a moon with the diameter of 400 km, so this means that Mimas the 21st largest moon in the solar system.

Is Mimas that far from Saturn?

Mimas is closer than you think. It is 186,000 km far from Saturn this distance is closer than the Earth and the moon to each other.

How buff is Mimas really?

This moon is weak and by weak I mean very WEAK, because this little moon only has 0.6% of Earth's gravity so you don't want to fall there or even jump!

What is Mimas made of?

To start of Mimas has the mean density of only 1.15 of water, aliens wouldn't like to go there, but since doesn't have a lot of water it is believed that Mimas is a composed principally of ice, let's not forget to mention that Mimas is very bright, by reflecting more than 80% of sunlight falling on it.

Does Mimas have an atmosphere?

Sadly Mimas is one of the moons to not have an atmosphere or even a magnetic field.

Is life possible on Mimas?

Scientists say that life on Mimas might be possible, because of Mimas's ocean, even if the ocean is encased in ice.

Mimas Facts:
Mimas is special for its largest crater Herschel.

The largest crater of Mimas has an amazing high mountain at the center of it.

Mimas was discovered back in September 17th, 1789.

This moon is believed to have created the Cassini Division.

Saturn/Rhea

Rhea is the second largest moon of Saturn just after Titan. Rhea's surface is comparable to Australia. Rhea is a small, airless body and cold moon.

How big is Rhea?

As we said Rhea is the second largest moon of Saturn and the ninth in the solar system, Rhea has the mean diameter of 1,527.6 km which is the third of Titans diameter!

How far is Rhea from Saturn?

Well Rhea is a bit far if we really look at it. Rhea is 527,000 km far from Saturn so in perspective it is farther than Tethys and Dione. Now because of this little thing Rhea does not receive at all! Ample tidal variation from its mother Saturn to cause internal heating, this is an important effect!

Does Rhea have an atmosphere?

Rhea has an atmosphere, but it is so thin! Scientists call Rhea atmosphere an exosphere. Rhea's atmosphere might be thin, but still has something not every object has, which is oxygen it also has carbon dioxide. Rhea's atmosphere was discovered back in 2010 by the Cassini space craft.

Is life possible there?

Unfortunately, even though Rhea has oxygen, but the oxygen there is not enough for humans, while the carbon dioxide could be caused by existence of rudimentary life? If life was possible on Rhea that would've been cool.

How buff is Rhea?

Rhea is a really weak moon since on there if you were to fall or jump you would instantly regret it, since the gravity on Rhea is only 0.264 m/s2. I never knew a moon would be this weak!

Rhea Facts:

Did you know the original name of Rhea was Saturn V since it was the fifth moon of Saturn?

Rhea has lots of water ice on its surface, which makes the moon highly reflective!

Rhea has 2 big craters that are on the side that faces the parent. These 2 craters are incredibly old and are from 400-500 km across.

Saturn/Tethys

Tethys is mid-sized moon which is great, Tethys has another name, which is Saturn lll it's called Saturn lll, because it's the third moon of Saturn and the fifth largest moon between Saturn's moons.

How big is Tethys?
Tethys is the fifth largest moon between Saturn's moons and the 16th in the solar system. Tethys has the mean diameter of only 1,062 km which is alright if we compare it to the microscopic moons.

Does Tethys have an atmosphere?
Tethys seems to be an unlucky moon; Tethys is one of the moons to not have an atmosphere, only some lucky moons get to have an atmosphere even if it was an exosphere.

How strong is Tethys?

Oh boy, I guess all Saturn's moons are weak since on Tethys the gravity force is only 0.145 m/s2, so you wouldn't like to stay for hours in the air, because you either fell or jumped!

Is there life on Tethys?

Tethys may not support life since it doesn't have oxygen (it doesn't have oxygen because of its atmosphere oh wait it has none), but Tethys has water, but not any water its ice water!

Tethys Facts:

It is 4.56 billion years old.

It was discovered back in 1684, it was one of the first moons to that we discovered from Saturn.

It has a very large crater on its surface called Odysseus.

It takes Tethys to orbit around Saturn a short time, which is only 45.3 hours!

Uranus

Uranus is the 7th closest planet to the sun and the third in size. The temperature on Uranus reaches to minus 200 Celsius degrees; also the rotational axis of Uranus is sideway. It is exception case for Uranus only is our solar system.

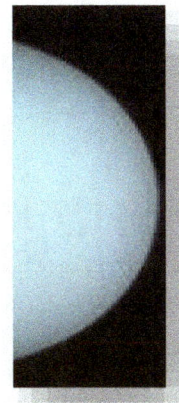

What are its moons like?

Uranus is known to have many moons and the biggest moon is called Titania, while the smallest is called Miranda. Every planet has different numbers of moons except for Mercury and Venus because they have none.

What are its rings like?

Uranus is known to have 13 small rings and the Inner rings are so dark so nobody can see them.

How many spacecrafts made it there?
Only one, yes only one, Voyager 2 is the only spacecraft that made it there, which in my opinion Uranus is a bit lonely.

Why Uranus is very cold? Because its surface form of liquid hydrogen and under this surface there is a layer of ice, and the center is a rocky ice core. Furthermore, the atmosphere is filled with ammonia and methane.

How big is Uranus?
Uranus is about four times wider than Earth. If Earth was a large apple, Uranus would be the size of a basketball.

What is the surface of Uranus like?
So Uranus has no true surface, but why? Since Uranus is an ice giant it might not have any surface so let's say a space craft went to

explore Uranus on the inside it wouldn't be able to land or even fly through it due to Uranus's crazy temperatures, because a metal space craft would be immediately destroyed!

How strong is it?

Uranus is going to shock you, since Uranus' gravity is a little less than here on Earth! The reason of that comes from the gravitational force, which depend on mass for Uranus (the solid mass), in other words the nature of liquid planet so its moment of inertia is too small.

Uranus facts:

It's the third biggest.

It has 27 beautiful moons.

It's a very cold planet in the solar system.

Uranus has blue tend to green color due to methane that consist in its atmosphere, where the reflection for the sun rays make this color appear.

Uranus is a cool planet rotating at nearly 90o it's like rotating on its side.

The rotational period around itself is 0.378 days, and rotates around the Sun in 84.32 years. Because its atmosphere (is filled with Ammonia, and Methane).

Uranus/Titania

Titania is the largest moon of Uranus and the first discovered between Uranus's moons with its brother Oberon. This moon is the eighth largest moon in the solar system.

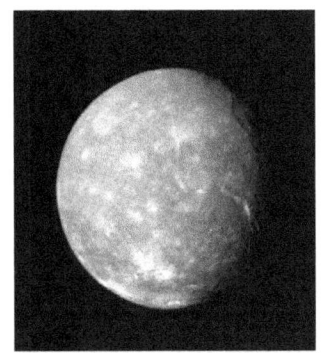

How big is Titania?

Since Titania is the eighth biggest it must mean it's kind of big right? Right Titania has the mean diameter of 1,578 km, which is crazy for a moon even though it's not a lot. Titania has the land surface like Australia, which is good!

Does Titania have an atmosphere?

Titania is lucky, because Titania has an atmosphere yay (probably, but yet not confirmed)! Titania has an atmosphere made of: Carbon dioxide (CO_2) like Callisto, but what about other gases? Well, my friends since

Titania is weak, so for example nitrogen or methane cannot be with Titania since Titania could not prevent them from escaping into space.

What are the names of Titania's sisters or brothers?

So, Titania has lots of sisters or brothers but the one closest to Titania by size is Oberon, but now let's go with the alphabet (from a to z): Ariel, Belinda, Bianca, Caliban, Cordelia, Cressida, Cupid, Desdemona, Ferdinand, Francisco, Juliet, Mab, Margaret, Miranda, Oberon (the closest to Titania), Ophelia, Perdita, Portia, Prospero, Puck, Rosalind, Setebos, Stephano, Sycorax, Titania (the one we are talking about), Trinculo, last but not least Umbriel. Oh boy these are lots of moons.

What is Titania's surface like?

Titania is an interesting moon of Uranus, since Titania's surface features rage from some impact craters to rift valleys and faults. There are some large impact basins on the surface. Most of Titania's craters are small.

Titania Facts:
It was discovered in 1787 by a British astronomer William Herschel.

The largest crater on Titania is named Gertrude.

Titania's largest crater was named after Hamlet's mother!

A year there is only 8.7 days pretty short in perspective every like a weak on Earth they celebrate a new year!

Its surface temperature about minus 220 Celsius.

It is difficult to study from terrestrial telescopes, so it is the only planet in the solar system that has been detected through mathematical equations rather than regular observations.

Neptune

Neptune is the 8th closest to the sun or we can say the 1st furthest from the sun, Neptune is the fourth largest planet. Neptune has at least five main rings and four more rings, which are clumps of dust likely formed by the gravity of a nearby moon. Now let's talk about Neptune more! In 2010 Neptune completed its first orbit, which is 165 years, since it was discovered back in 1846.

What are its moons like?
Neptune is known to have 14 beautiful moons and the biggest of them is: Triton. Triton is such a beautiful moon that I love with all of my heart!

What is Neptune's surface like?
Neptune doesn't have a solid surface. Its atmosphere, which is mostly made of

hydrogen, helium, and methane, extends to great depths.

How big is Neptune?

Neptune is so big its 49,244 km which is 4x the diameter of Earth!

How strong is Neptune?

Neptune is stronger than Earth, since the gravity force is 11.15 m/s2, which is 1.14 times stronger than Earth!

Neptune Facts:

Neptune's atmosphere is thick and windy.

Six rings surround Neptune.

A year on Neptune is longer than a century that's because a year there is 165 earth years, now that's a really long year.

Sadly, only voyager 2 visited Neptune.

Sometimes it's the furthest planet from the sun.

It's the only planet we can't see with our naked eye.

It is 30× the distance from the earth to the sun.

Neptune/Triton

Triton as we all know is the largest satellite of Neptune; it was the first moon of Neptune to be discovered in 1846 on October 11 by the English astronomer William Lassell. Fun fact about Triton (don't tell anyone I told you): Triton is the only moon in the solar system to orbit on the opposite side of its mother!

How big is Triton?

Triton is big it has the mean diameter of 2,700 km. Its surface is 23 million km2 which is 4.5% of Earth which is good or we could say 15.5% of Earths land area. This all means that Triton is the 7th biggest moon in the solar system.

Does Triton have an atmosphere?

Triton does have an atmosphere yay! But Triton's atmosphere is thin, but it is made of nitrogen like Earth with some really small amounts of methane, but how did Triton get to have an atmosphere? It is believed that Triton volcanic activity, which is driven by seasonal heating by the Sun!

What is Triton made of?

Triton is both interesting and common. Triton has the amazing core that is made of rock and metal and a beautiful crust made of frozen nitrogen over an icy mantle to cover the core. Fact about Triton: Triton has twice the density of water; this is higher density that any measured any other satellite almost in the outer planets, but Europa and Io have higher densities they don't like being defeated.

Is Triton buff?

Triton is kind of buff and kind of not since Triton has 8% of Earth's gravity. Triton has the gravity pull of 0.779 m/s2 so you are not going to stay for hours in the air, but you are probably going to take some time till you land.

What are Triton sisters or brothers named?

So, we said Triton has 14 sisters or brothers (as you like) we are going to say them in order with the alphabet (a to z): starting off with Despina, then on the list we have Galatea we skipped lots of letters, now let's move on with Halimede this moon was discovered back in 2002, then we got Hippocamp pretty funny name right there, then we have a beautiful moon named Laomedeia lots of astronomers discovered this moon, next on our list we have Larissa this moon it was back in discovered in 1981, now we have Nereid this moon was discovered by voyager 2 back in 1989, next on the beautiful list we have Nereid just like nerd but this moon was discovered back in 1949 pretty cool date there, now the last one with the letter N is Neso this moon was discovered back in 2002, now let's skip M and move to P with Proteus this beautiful moon was discovered in 1989, now we have Psamathe this amazing moon was discovered in 2003, now we have the letter S with Sao this gorgeous moon was discovered in 2002 by some lots of astronomers, we are getting close to the end, since we have the last two that are Thalassa this moon was discovered in 1989, then finally we have Triton.

Triton Facts:

Triton might crash into Neptune in the future.

Triton might got pulled from the Kuiper belt to get to be in Neptune's orbit.

Triton has ice volcanos which is cool.

Triton is just slightly smaller than our moon.

Triton is thought to have a young surface compared to the solar systems age.

Pluto

If Pluto was considered a planet in the solar system it would be the coldest planet. Pluto's temperatures can go down till minus 240 Celsius degrees, you're going to turn into an ice cube in 0.001 seconds when you enter Pluto due to its crazy temperatures.

What are its moons like?
Even though Pluto is the smallest planet but it has more moons than the earth coming up to 5 and the biggest is Charon. Pluto and Charon are considered a double planet while orbiting each other, since Charon is half the size of Pluto.

How big is Pluto?
Pluto is so small, smaller than our moon, Pluto has only two thirds of the moons diameter, since Pluto's diameter is 2,370 km, which is 1\6th the size of Earth!
What are its rings like?

Pluto is the only planet from the outer planets that doesn't have a ring.

Pluto Facts:

A day there is long since its 153 hours so if you think you don't have enough time here on Earth you might want to go to Pluto.

A year there is longer than two centuries since its 248 Earth years

It's a dwarf planet

It was accidentally discovered by an astronaut named: Clyde Tombaugh back in 1930!

It's sometimes the 9th planet in the solar system.

Its atmosphere is filled with frozen Ammonia and Methane.

The discovery of Pluto followed on from that of Neptune. Neither the orbits of Pluto nor Neptune were well defined (overlap), but it was thought that there might be a more distant planet called 'planet X'.

In 2006 astronomers considered it as a Dwarf planet.

Glossary

Asteroids: The minor planets, most of which move around the Sun between the orbits of Mars and Jupiter. Several thousands of asteroids are known; much the largest is Ceres, whose diameter is 1003 km. Only one asteroid (Vesta) is ever visible with the naked eye.

Astronomer: a person who studies space.

Aurora: a phenomenon which appears in the polar region in earth, in the form of colored curtains for minutes or hours. The reason is that when the charged electrons that comes from sun winds interact with the upper gasses in the earth atmosphere, so that the the radiation with different colors occurs in the sky.

Comet: celestial objects that shows a distinct luster from a head extending from a long tail in the form of luminous cloud that moving around the Sun in an orbit.

Crater: A hole on a planet or a moon which a meteorite creates.

Core: the very right center of a planet or a star.

Density: The mass of a substance per unit volume. Taking water as one, the density of the Earth is 5.5.

Dwarf planets: celestial bodies that revolves around the sun, which have enough mass to overcome the force of gravity on the strength of the body to become in small spherical shape.

Ecliptic: defined as 'the apparent yearly path of the Sun against the stars', passing through the constellations of the Zodiac. Since the plane of the Earth's orbit is inclined to the equator by 23.5 degrees, the angle between the ecliptic and the celestial equator must also be 23.5 degrees.

Fire Arches: Streams of ionized gas emanating from the surface of the Sun, also their arc shape is controlled by the strong magnetic field. They last for minutes or hours, and some of them are thrown away into space.

Galaxy: system of stars, as an example our Milky Way contains about 100000 million stars, but is not exceptional in size. They are in many different shapes. Some are spiral, elliptical, and irregular.

Great Red Spot: is the one of the most famous features of Jupiter. It is a continuous hurricane. Some mathematical models suggest that this storm is a permanent feature of the planet. Because of the magnitude of this storm, which can be observed from the ground using A14 cm telescope of larger.

Greenhouse Effect: The increase in carbon dioxide in the atmosphere, as the visible radiation penetrates the atmosphere and interact with the soil and return into infrared radiation, which does not allow carbon dioxide to escape into space and remain trapped, as a result leads to high temperature on earth.

Inner Planets: Mercury, Venus, Earth, and Mars. These planets do not have surround rings also, they have few moons and consist mainly of rock and mineral materials.

Kuiper Belt: A region of frozen objects and rocks orbiting the Sun, outside the orbit of Pluto, and it is the source of comets. This region is away from the Sun about 100 AU.

Lava: the molten rock above the planet's surface.

Mass: it is a measure of inertia for an object.

Meteor (Cometary debris): a small particle which enters the Earth's upper atmosphere and burns away, producing the effect known as a shooting star (meteor showers).

Meteoroids: A larger body, which is able to reach ground level without being destroyed. They more nearly related to an asteroid or minor planet. Meteorites may be stony, iron or of intermediate type. In a few cases meteorites have produced craters.

Nebula: a mass of tenuous gas in space together with what is loosely termed 'dust'. If there are stars in or very near the nebula, the gas and dust will become visible, either because of straightforward reflection or because the stellar radiation excites the material to self-luminosity.

If there are no suitable stars, the nebula will remain dark, and will betray its presence only because it will blot out the light of stars lying beyond it. Nebulae are regarded as regions in which fresh stars are being formed out of the interstellar material.

Nebular theory: it explains the formation and evolution of the solar system. It is part of the Big Bang theory. This theory was reached by astronomers in the last decades of the 20th century.

Outer Planets: Jupiter, Saturn, Uranus, and Neptune. These planets are characterized by ring systems, they have a large mass, they have numerous natural moons, and they have liquid surfaces and revolve around itself quickly.

Planet: the celestial object that is stronger by every object that orbits the star by mass.

Rings: things that form a circle around a planet that are mostly made of broken moons.

Solar Explosions: Heavy explosions of ionized gas from the surface of the Sun give a large amount of x-rays, it also damages the communications and the technology of the satellite navigation signals.

Star: A big object in space that produces light itself.

Sunspots: spots on the surface of the Sun appear as dark spots because they are cooler than the average surface temperature.

Terrestrial planets: the four closest planets to the sun (Mercury, Venus, Earth, Mars).

Latest space news

Elon Musk just lost 14 billion dollars while testing the biggest space rocket ever made and it only lasted 4 min. which is a great success.

NASA went back to the moon to explore more things.

We discovered that the Milky Way galaxy (which is our galaxy) and the Andromeda galaxy are going to collide in 5 billion years.

NASA is willing to make cloud cities on Venus.

Elon Musk says that the first city on Mars will be in glass dorms.

The closest black hole to us is only 1,550 light years away. I know that sounds really far but it is not when you scale it on the Milky Way galaxy map.

NASA is willing to launch the Psyche spacecraft.

Bibliography

Bartels, M. (2020, September 14). The phosphine discovered in Venus' clouds may be a big deal. Here's what you need to know. Space.com.

Batygin, K., & Brown, M. (2016). Evidence for a distant giant planet in the solar system. The Astronomical Journal, 151(2).

Charles Q. (2023, March 17). Saturn: Everything you need to know about the sixth planet from the sun. Space.com.

Choi, C. (2021, October 1). Mars: What we know about the Red Planet. Space.com.

Choi, C. (2017, May 12). Planet Neptune: Facts about its orbit, moons & rings. Space.com.

Choi, C. (2019, July 10). Uranus: The ringed planet that sits on its side. Space.com.

Choi, C., & Dutfield, S. (2021, October 31). Planet Mercury: Facts about the planet closest to the sun. Space.com.

Choi, C., & Dutfield, S. (2022, January 26). Saturn: Facts about the ringed planet. Space.com.

Choi, C., & Gohd, C. (2021, August 16). Venus: The hot, hellish & volcanic planet. Space.com.

Horton, J. (2021, November 18). Is there water on Mars? Live Science.

Howell, E. (2016, June 30). Neptune's moons: 14 discovered so far. Space.com.

Kornreich, D. (2019, January 28). How long does it take for the sun's light to reach us? Ask an Astronomer, Cornell University.

Letzer, R. (2020, June 4). Mars once had rings and a much bigger moon, new evidence suggests. Live Science.

NASA (2023, July 21). Inner solar system.

NASA Goddard Space Flight Center. (2020, October 22). The Milky Way.

NASA Science. (2021, October 30). Jupiter.

NASA Science. (2021, August 30). Our solar system.

NASA Science. (2023, July 21). Outer solar system.

NASA Science. (n.d.). Saturn moons.

NASA Science. (2023, July 20). Small Bodies of the Solar System

NASA Science. (2017, November 30) Kuiper Belt: In Depth

Night Sky Network. (2017, June). Solar system, galaxy, universe: What's the difference? NASA Jet Propulsion Laboratory.

Pultarova, T. (2021, June 30). No hope for life in Venus clouds. Live Science.

Specktor, B. (2019, May 6). Scientists think they've found the ancient neutron star crash that showered our solar system in gold. Live Science.

Specktor, B. (2018, October 1) Pluto should be a planet and so should Earth's moon, new study claims. Live Science.

Tillman, N. T. (2021, August 16). Multiple supernovas may have implanted our solar system with the seeds of planets. Space.com.

Tillman, N. T., & Dutfield, S. (2022, January 17). How was the moon formed? Space.com.

Tobias Chant Owen, (2023, June 6). Solar system astronomy.britannica.com.

University of California, San Diego. (2002). The Jovian planets.

U.S. Geological Survey. (2019, November 13). How much water is there on Earth?

Wall, M. (2020, February 25). Alien-life hunters are eyeing icy ocean moons Europa and Enceladus. Space.com.

NASA. Mars fact sheet. NASA. Retrieved July 11, 2022, from www.nssdc.gsfc.nasa.gov/planetary/factsheet/marsfact.html

NASA. Mars. NASA. Retrieved July 11, 2022, from www.solarsystem.nasa.gov/planets/mars/overview/NASA. NASA Mars Exploration. NASA. Retrieved July 11, 2022, from https://mars.nasa.gov/

NASA. Send your name to Mars. NASA. Retrieved July 11, 2022, from www.mars.nasa.gov/participate/send-your-name/future

US Department of Commerce, N. O. A. A. The planet Mars. National Weather Service. Retrieved July 11, 2022, from www.weather.gov/fsd/mars

Erick J. Cano et al, "Distinct oxygen isotope compositions of the Earth and Moon", Nature Geoscience, Volume 13, March 2020, https://doi.org/10.1038/s41561-020-0550-0

Raluca Rufu, "A multiple-impact origin for the Moon", Nature Geoscience, Volume 10, January 2017, https://doi.org/10.1038/ngeo2866

Edward Belbruno et al, "Where Did the Moon Come From?", The Astronomical Journal, Volume 129, March 2005.

Thomas S. Kruijer and Gregory Archer, "No 182W evidence for early Moon formation", Nature Geoscience, Volume 14, September 2021, https://doi.org/10.1038/s41561-021-00820-2

Index

Introduction	3
Solar system	5
The Sun	9
Mercury	17
Venus	22
Earth	28
Moon	39
Mars	45
Mars/Phobos	52
Mars/Deimos	54
Asteroid Belt	55
Ceres	59
Jupiter	62
Jupiter/Io	68
Jupiter/Ganymede	72
Jupiter/Europa	76
Jupiter/Callisto	80

Saturn	84
Saturn/Titan	87
Saturn/Mimas	91
Saturn/Rhea	94
Saturn/Tethys	97
Uranus	99
Uranus/Titania	103
Neptune	106
Neptune/Triton	108
Pluto	112
Glossary	114
Latest space news	120
Bibliography	121
Index	125

www.ingramcontent.com/pod-product-compliance
Lightning Source LLC
Chambersburg PA
CBHW050307230526
45471CB00005B/2071